Types of Plants

Printed in Mexico

ISBN-13: 978-0-15-362022-5

ISBN-10: 0-15-362022-6

6 7 8 9 10 0908 16 15 14 13 12

4500358761

Visit *The Learning Site!*
www.harcourtschool.com

Lesson 1

What Do Plants Need to Live?

VOCABULARY
roots
nutrients
stem
leaf

Roots are the parts of a plant that grow underground. Roots take in water and nutrients from the soil.

Nutrients are things in the soil that help plants grow and stay healthy.

A **stem** is the part of a plant that grows above ground. It helps hold up a plant. This cactus has a thick stem.

A **leaf** is the part of a plant that grows out of a stem. It is where a plant makes its food.

READING FOCUS SKILL

MAIN IDEA AND DETAILS

The main idea is what the text is mostly about. Details tell more about the main idea.

Look for details about the things that plants need to live.

What Plants Need

Plants are living things. They grow in many places. But most plants need the same things to live. They need soil, water, air, and light. Plants cannot live if these needs are not met.

People care for some plants. But most plants grow without help. They get what they need from the sun, the air, the rain, and the soil.

Name four things plants need to live.

This gardener helps plants get air, light, water, and soil.

Roots and Stems

A plant has parts that help it get what it needs to live. **Roots** are plant parts that grow underground. Roots help hold a plant in the soil. They also take in water and nutrients from the soil. **Nutrients** are things in soil that help plants grow and stay healthy.

A **stem** is a plant part that grows above ground. It holds a plant up. A stem carries water and nutrients from the roots to the leaves. It also carries food from the leaves to the roots.

Focus Skill **Tell how water from the soil travels to the leaves of a plant.**

Leaves

A **leaf** is a plant part that grows out of a stem. Leaves are where a plant makes its food. Leaves use sunlight, air, and water to make food. This food helps the plant live and grow.

Focus Skill **What do leaves use to make food?**

How Plants Live in Different Environments

Plants live almost everywhere. The plant parts help them get what they need.

A cactus lives in a desert. It has a thick stem to store water. A water lily grows in water. It has a long stem so its leaves can float on top of the water to get sunlight.

Focus Skill **Why do a cactus and a water lily have different stems?**

▲ **Cactus**

Review

Focus Skill **Complete this main idea statement.**

1. Most plants need soil, ______ , air, and light to live.

Complete these detail sentences.

2. ______ hold a plant in soil and take in water and nutrients.
3. A plant's ______ holds a plant up.
4. A plant's leaves make ______.

Lesson 2

What Are Some Types of Plants?

VOCABULARY

seeds
deciduous
evergreen

Seeds are the first stage of life for many plants. Each pea is a seed that could grow into a new plant.

Deciduous plants lose all their leaves at the same time each year. Many Oak trees are deciduous.

An **evergreen** plant stays green and makes food all year long. Magnolia trees are evergreens.

READING FOCUS SKILL
MAIN IDEA AND DETAILS

The main idea is what the text is mostly about. Details tell more about the main idea.

Look for details about different kinds of plants.

Grouping Plants

Scientists group plants by putting plants that are alike together. Most plants have seeds. **Seeds** are the first stage of life for many plants. Scientists put plants with seeds in two groups. One group has flowers. The other group does not. All the plants on this page have seeds and flowers.

Tell why grouping plants is helpful.

These plants all have seeds and flowers.

◀ Maple leaf

Palmetto leaves ▶

Types of Leaves

Leaves come in many shapes and sizes. Scientists group plants by the kind of leaves they have. They also group plants by how they lose their leaves.

Deciduous plants lose all their leaves at the same time each year. This often happens in the fall. Most oak trees are deciduous.

Evergreen plants stay green all year. They lose leaves or needles from time to time. But they do not lose them all at once. Pine trees are evergreens.

How are leaves used to group plants?

◀ Oak tree

pine needles ▶

Flowers

Scientists also group plants by whether they have flowers. Flowers have special parts that make seeds. New plants can grow from these seeds.

Plants without flowers make seeds in other ways. Many evergreens make seeds inside their cones. Other plants make seedlike parts called spores. Like seeds, spores are the first stage of life for new plants.

Tell how plants without flowers make new plants.

The flowers on this cherry tree can make seeds. ▶

▲ This fern makes spores.

Seeds

Seeds from different plants are different shapes, sizes, and colors. But the seeds are alike in important ways.

Each seed has food for a new plant to grow. Each seed looks different from the plant it will become. But each seed grows into a plant like the one that it came from.

Tell how all seeds are alike.

Orange seeds can grow into orange trees. ▶

Focus Skill **Complete this main idea sentence.**

1. ______ can be grouped in many ways.

Complete these detail statements.

2. ______ are the first stage of life for many plants.
3. Plants that lose all their leaves at the same time are called ______ .
4. ______ plants keep their leaves all year long.

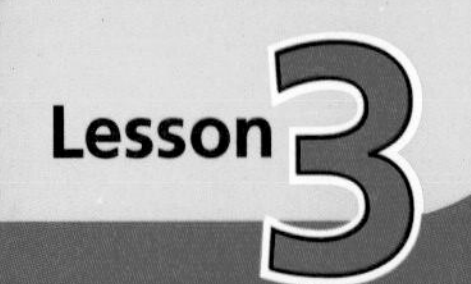

Lesson 3

VOCABULARY

photosynthesis
chlorophyll

How Do Plants Make Food?

Photosynthesis is the process that plants use to make food.

Chlorophyll is a green substance in leaves. It helps plants use light energy to make food.

READING FOCUS SKILL

CAUSE AND EFFECT

A **cause** is what makes something happen. An **effect** is what happens.

Look for the **effect** of plants using water, light, and carbon dioxide.

How Plants Get Energy

All plants make their own food. They use this food for energy. The food plants make is sugar. **Photosynthesis** is the process plants use to make sugar.

Tell what plants use photosynthesis to make.

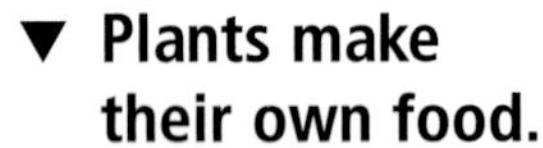

▼ **Plants make their own food.**

Photosynthesis

During photosynthesis, plants make sugar by combining water and a gas called carbon dioxide. The plant gets energy from sunlight to make the sugar.

Chlorophyll is a green substance inside leaves. It helps the plant use light energy to make its food.

What is the effect of chlorophyll?

How Plants are Helpful and Harmful

Plants are helpful in many ways. People eat plants for food. This gives us energy.

People also use the parts of plants to build and make things. We use wood to build houses and make paper. We use some plants to make cloth and medicines.

People get energy when they eat strawberries.

Grains feed many people.

Some plants can be harmful. They may have poison in their parts. Some plants can cause allergies.

Tell how plants can be helpful and harmful.

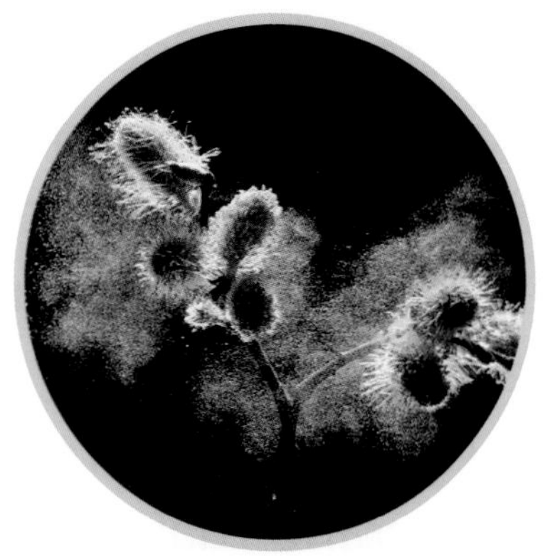

▲ Pollen makes some people sneeze and cough.

◀ Poison ivy makes some people itch.

Review

Focus Skill

Complete these cause and effect statements.

1. In the process of ______ plants turn water and carbon dioxide into sugar.
2. A plant uses light energy and ______ to make food.
3. People eat plants to get ______.
4. Some harmful plants can cause ______.

GLOSSARY

chlorophyll (KLAWR•uh•fihl) The green substance inside leaves that helps a plant use light energy to make its food (17)

deciduous (dee•SIJ•oo•uhs) Relating to plants that lose all their leaves at the same time every year (11)

evergreen (EV•uhr•green) A plant that stays green and makes food all year long (11)

leaf (LEEF) The part of a plant that grows out of the stem and is where the plant makes food (6)

nutrients (noo•tree•uhntz) Things in the soil that help plants grow and stay healthy (5)

photosynthesis (foht•oh•SIHN•thuh•sis) The process that plants use to make sugar (16)

roots (ROOTS) The parts of a plant that grow underground and take water and nutrients from the soil (5)

seeds (SEEDZ) The first stage of life for many plants (10)

stem (STEM) The part of a plant that grows above ground and helps hold the plant up (5)